Paratheory of Gravity

Truth Revealed
By
Elmer Para

Elmer Para

Copyright © 2024 Elmer Para

All rights reserved. No part of this book may be reproduced, stored in a retrieval system, or transmitted in any form or by any means-electronic, mechanical, photocopying, recording, or otherwise without the author's prior written permission.

Disclaimer

The Paratheory of Gravity theory presented in this book is a proposed explanation for the universe's workings. The author acknowledges that further research and testing are required to validate the claims. The views expressed are solely those of the author and do not necessarily reflect those of affiliated institutions or organizations. The author disclaims any responsibility for using or interpreting the information provided in this book.

Elmer Para

Table of Contents

PARATHEORY OF GRAVITY ------------------------ **0**

Truth Revealed -- **0**

Elmer Para --- **0**

Copyright © 2024 Elmer Para --------------------------------- **1**

Disclaimer --- **1**

INTRODUCTION -------------------------------------- **5**

SECTION 1 --- **6**

DEFINITION OF GRAVITY ------------------------ **6**

Spacetime -- **7**
   Experiment A: Spacetime Compression ------------------------- 8

EXPERIMENT B: THE PARABLE: A DENSELY CONCENTRATED REGION OF SPACETIME AT THE CORE -- **9**
   Experiment C: Spacetime Compression and Gravitational Propagations: Highlighting the limitations of both Einsteinian and Newtonian gravity ------------------------------ 10

Experiment D: Spacetime Compression and Gravitational Convergence: Inadequacy of both Einsteinian and Newtonian Gravity --- 12
Experiment E: Gravity's Effects in a Closed Room Kilometers Deep Underground --- 13
Experiment F: Unveiling the Flaws and Limitations of Albert Einstein's Theory of Gravity --- 15
Experiment G: Extended Gravitational Free-Fall Experiment Revealing limitations of both Albert Einstein's and Isaac Newton's theories of Gravity --- 17

SECTION 2 --- 20

THE COSMIC METAMORPHOSIS --- 20

The Astronaut in Outer Space --- 20
From floating astronaut to floating Earth --- 20
The Celestial Body's Protective Mechanisms --- 22
Reaching the Size of the Sun: From Localized to Universal Gravity --- 25
Pure Energy --- 26
Interconnected Solar System --- 27
The Dynamic Interplay of the Sun and Earth --- 28
Equating black holes as anti-matter --- 29
A star billion times larger than the Sun: The Explosive Transformation of the Parable into a Black Hole --- 30
The birth of a black hole --- 31
The Dance and Sway of Black Holes --- 31
The Final Image --- 32

SECTION 3 -- **34**

INFINITE SELF-CREATING UNIVERSE -------- **34**
   Illusion of Expansion --34
   Genesis of Creation ---36
   Conclusion --37

Introduction

In the quest to unveil the universe's beauty, countless thinkers, philosophers, and scientists have embarked on a relentless journey. While the tales of philosophers and religious figures intrigue us, scientists delve deep into complex mathematics. Yet, despite their best efforts, we find ourselves guided by an invisible force, leaving the entire workings of gravity just beyond our grasp. I humbly present my revolutionary theory of gravity, aiming to inspire minds and challenge the limitations of existing theories. Let us elevate our understanding of the cosmos and partake in this enlightening and enriching exploration.

Could this be the theory that finally unlocks the secrets of the universe? Open these pages and decide for yourself.

Section 1

Definition of Gravity

Gravity is a simultaneous reaction of spacetime to the presence of an object. This reaction is localized, spherical inward compression of spacetime that passes through all atoms, imparting weight, and converges at the core, creating a densely concentrated region of spacetime, which I coin as a 'parable.' This localized spacetime compression from all directions enables celestial bodies to float, exhibit equal weight distribution across their surface, maintain roundness, and possess an organized internal structure.

This spherical inward compression of spacetime is analogous to a sponge compressing evenly around an inserted round object. The key difference is that the spherical inward spacetime compression passes through all atoms, weakening the compression. What remains at the Earth's surface is what we perceive as gravity.

Matter and gravity are intimately intertwined; one cannot exist without the other in forming celestial objects.

Spacetime

Space is infinite and cannot be contained by any finite means. It facilitates everything, and Anything that occupies it is governed by the rule of time. Space and time combine to make "spacetime" an infinite medium, a flawless, indestructible structure, creator, and container of everything. Finite materials such as galaxies, stars, planets, and all other observable objects are merely creations of spacetime, together forming the universe itself.

Since gravity is fully functional on our planet, Earth is an ideal testing ground for various theories of gravity, including those proposed by Albert Einstein and Isaac Newton. Any theories that cannot accurately explain the gravitational phenomena on Earth, their applicability and validity on a universal scale must be questioned and scrutinized. This rigorous examination is necessary to pursue scientific truth to advance our understanding of the universe.

This flowing-inward-like sensation that pushes skydivers toward the ground is a manifestation of spacetime compression in action, which we interpret as gravity. To better understand this concept, consider the following thought experiments involving balls and the Earth:

Experiment A: Spacetime Compression

Imagine three iron balls simultaneously entering tunnels at varying angles (30°, 40°, and 50°) originating from a single point on Earth. These balls will emerge at the other side at the same time, regardless of the angles. The equation "time equals distance" can encapsulate this phenomenon, where time represents the angles and distance represents space. This experiment demonstrates the seamless interplay between time and space guiding the balls through the tunnels, confirming that gravity is a spacetime compression.

Experiment B: The Parable: A Densely Concentrated Region of Spacetime at the Core

Objective:

To demonstrate the existence of a densely concentrated region of spacetime at the Earth's core that acts as an impact point, preventing the bowling ball from passing through the core even without matter in the tunnel.

Setup:

Imagine a straight tunnel passing through the Earth's core vertically or horizontally with a bowling ball drawn into the tunnel.

Prediction:

According to the spacetime compression theory, the bowling ball would become trapped at the parable, a region where spacetime is highly concentrated. This dense spacetime concentration acts as an impact point preventing the bowling ball from passing through the Earth's core, regardless of the absence of matter in the tunnel.

Experiment C: Spacetime Compression and Gravitational Propagations: Highlighting the limitations of both Einsteinian and Newtonian gravity

Imagine a 45° angled tunnel passing through the Earth. Release a bowling ball into this tunnel. According to my theory, the spherical inward compression of spacetime causes the ball to propagate and pass through the center while maintaining its momentum as it moves toward the opposite side. However, the spacetime compression from the opposite side counteracts the ball's momentum, causing it to oscillate back and forth continuously before gradually coming to rest.

Applying Albert Einstein's concept of spacetime curvature to this experiment, the bowling ball would not propagate due to the absence of matter required to curve the spacetime inside the tunnel. Preventing the ball from propagating. A curvature concept cannot influence the ball's propagation back and forth at the tunnel.

According to Einstein's theory, the presence of matter is required to curve spacetime and create the gravitational effects we observe in the universe.

Similarly, applying Isaac Newton's law of gravitation to this experiment would cause the bowling ball to stick to the tunnel's wall due to the attraction between the ball and the tunnel wall, thus preventing the ball from propagating through the tunnel. This simple experiment highlights the limitations of Einstein's and Newton's theories of gravity.

Elmer Para

Experiment D: Spacetime Compression and Gravitational Convergence: Inadequacy of both Einsteinian and Newtonian Gravity

Imagine a cross-earth tunnel drawing four bowling balls weighing 10 kilograms each. According to my theory, the spherical inward compression of spacetime that passes through imparts weight and converges at the core. These balls would converge at the core with a combined weight of 40 kg as self-created weight. Scaling this up to the size of the Earth, imagine the whole volume of the Earth converging at the core. This results in an organized internal structure, a round shape, and equal weight distribution across the planet. Due to this inward compression of spacetime from all directions, the Earth becomes both self-created and floating weight. Paradoxically, Earth floats weightlessly through space.

Albert Einstein's theory of relativity and Isaac Newton's law of gravitation cannot explain this phenomenon.

Experiment E: Gravity's Effects in a Closed Room Kilometers Deep Underground

Objective:

To compare the gravitational effects experienced by a person in a closed room located kilometers deep underground according to three different theories: spacetime compression theory, Einstein's curvature theory, and Newton's Law.

Setup:

Consider a closed room located kilometers deep underground. A person is inside this room.

Predictions Based on Three Different Theories:

1. **Spacetime compression theory**: according to this theory, the person in the room would still experience gravity due to the inward compression of spacetime that passes through and converges at the core.
2. **Einstein's curvature theory** states that a person would appear disoriented or floating due to the absence of matter required to curve spacetime inside the room.

3. **Newton's Law:** A person would be drawn toward the surrounding matter, causing them to stick to the wall, ceiling, or floor due to the attraction between the person and the walls.

Experiment F: Unveiling the Flaws and Limitations of Albert Einstein's Theory of Gravity

This experiment critically examines Albert Einstein's renowned analogy involving a heavy object, such as a planet or Star, placed on a flat sheet of spacetime to represent spacetime curvature. In this scenario, the experiment reveals a paradoxical outcome by replacing the curvature with a scale: the scale becomes heavier than the Earth at any given point on the planet.

The famous rubber sheet analogy involving the celestial body in a flat sheet is inconsistent because the heavenly body in question, my entire definition of gravity, is already at play. Einstein's curvature of spacetime is redundant or unnecessary. Removing the space-time curvature of the Sun would have no observable consequences; however, removing the spherical inward compression of spacetime around the Sun would disperse the Sun's matter from its spherical form.

Heaviness should be considered a localized experience arising from the localized, spherical inward compression of spacetime that passes through and converges at the core evenly from all directions. This enables the celestial body to float weightlessly through space. Challenging the concept of space-time curvature proposed by Albert Einstein.

Einstein's theory of relativity states that massive objects can cause spacetime to curve, introducing an outward force caused by weight or mass. However, celestial bodies floating in space contradict this. Because they are subjected to simultaneous spherical inward compression of spacetime, passing through and converging at the core evenly from all directions to keep them floating in space effortlessly. This phenomenon prevents them from falling in any specific direction.

Albert Einstein's theory of relativity, which proposes the concept of spacetime curvature, is fundamentally flawed as it fails to provide a comprehensive explanation for several critical aspects of celestial bodies. The theory cannot account for celestial bodies' roundness and organized internal structure. Nor can it explain the equal weight distribution across the planet. Moreover, Einstein's theory falls short of addressing the weightlessness experienced by celestial objects in space. Also, it falls short of addressing that gravity and matter are intertwined, and neither can exist without the other in the context of a celestial body, which is crucial for the formation and existence of these heavenly bodies.

Elmer Para

Experiment G: Extended Gravitational Free-Fall Experiment Revealing limitations of both Albert Einstein's and Isaac Newton's theories of Gravity

To further illustrate the principles of the spacetime compression theory. Consider the famous gravitational Free-Fall experiment. In this experiment, objects of different masses are dropped from the same height, and their acceleration towards the ground is observed. According to my spacetime compression theory, these objects in different masses will fall simultaneously or at the same rate due to the uniform spherical inward compression of spacetime passing through and converging at the core.

Additionally, upon hitting the ground, the constant passage of spacetime compression through the atoms of the objects of different compactness results in different perceived weights; atomic compactness affects weight because spacetime compression passes through all atoms, imparting more weight to denser objects.

Furthermore, the constant passage of spacetime compression guides matters inward, imparting weight and converging at the core from all directions, where it can no longer pass through itself. This creates an unimaginably dense self-created weight called the parable, amplifying the celestial body's gravitational ability, stability, and weightlessness in space.

Applying Isaac Newton's theory of gravity to the Gravitational free—fall experiment would predict that objects with different masses should fall at different rates due to the varying attractive force between them and the Earth.

According to Newton's theory, the gravitational force between two objects is directly proportional to the product of their masses and inversely proportional to the square distance between their centers of mass.

This relationship is presented by the equation $F = G * (m1 * m2) / r^2$, where G is the gravitational constant. Thus, a more massive object would experience a stronger attraction to the Earth, resulting in objects of different masses falling at different rates.

However, as demonstrated in the Gravitational Free-Fall Experiment, objects with different masses are found to fall at the same rate. This discrepancy highlights the limitations of Newton's theory. It emphasizes the need for alternative explanations, such as the spacetime compression theory, to describe objects' behavior in this classic Experiment accurately.

Regrettably, Albert Einstein's curvature theory of relativity cannot be applied to this Gravitational Free-Fall Experiment. The experiment cannot be conducted at any point on the planet because Einstein's concept of curvature fails to account for the equal weight distribution across the Earth's surface. This aspect is crucial for understanding why objects in different masses fall at the same rate during free fall.

The experiments discussed have effectively exposed the significant limitations and inconsistencies in Albert Einstein's and Isaac Newton's theories of gravity. Since these theories cannot accurately explain the gravitational phenomenon on our planet, their applicability and validity on a universal scale must be questioned and scrutinized in pursuit of scientific truth.

While their work has undeniably contributed to our understanding of the cosmos, it is crucial to acknowledge that these theories struggle to explain certain gravitational phenomena observed on Earth accurately.

Having established the limitations and inconsistencies of existing theories and laid the foundation for an alternative approach, let us now delve deeper into my proposed theory's core principles and implications in the following section through the cosmic metamorphosis.

Section 2

The Cosmic Metamorphosis

We will journey through cosmic metamorphosis, demonstrating the dynamic interplay between spacetime and objects of all sizes. We will explore various transformations, beginning with astronauts in outer space and transforming into the Earth to the Sun and a star a billion times larger than the Sun, ultimately into a black hole.

The Astronaut in Outer Space

Imagine a solitary astronaut floating in the boundless realm of outer space, experiencing the awe-inspiring sensation of weightlessness. This remarkable state arises because spacetime passes through all atoms of the astronaut's body from all directions, rendering them unable to swim, maneuver, or change their position. They are wholly suspended in this cosmic embrace.

From floating astronaut to floating Earth

As the astronaut expands to the size of the Earth, spacetime simultaneously compresses them inward from all directions, forming the astronaut into a sphere. This compression is analogous to a sponge compressing evenly around an inserted round object. The only difference is that spacetime compression passes through all atoms, which weakens the compression. What remains on the surface of the Earth is what we perceive as gravity, which we experience daily in our lives.

On this Earth, my entire definition of gravity is entirely in effect. Imagine a cross-Earth tunnel drawing four bowling bowls weighing 10 kg each. According to my theory, the spherical, inward compression of spacetime, these balls would converge at the core with a combined weight of 40 kilograms as self-created weight. Scaling this up to the size of the Earth, imagine the whole volume of the Earth converging at the core. This results in an organized internal structure, a round shape, and equal weight distribution across their surface. Moreover, due to the localized, spherical, inward compression of spacetime from all directions, Earth becomes both self-created and floating weight. Paradoxically, Earth floats weightlessly through space, riding in an invisible gravitational bubble (the spherical gravitational field). To better understand this invisible bubble, imagine spacetime is liquid. When it compresses inward in spherical form, it creates a simultaneous wave, a crown-like structure that flows sideways continuously, shielding the edges of the spherical gravitational field. If we throw an apple below the crown, it will be pushed back toward the Earth. If we throw an apple into the crown, it will orbit the planet.

The Celestial Body's Protective Mechanisms

Celestial bodies are encapsulated or riding in invisible gravitational bubbles or fields that act as protective gear. This phenomenon ensures that two heavenly bodies cannot simply collide. For instance, when Mercury moves closer to the sun, the gravitational bubbles of both celestial bodies interact, preventing a catastrophic collision and causing Mercury to speed up. When two gravitational bubbles are in proximity, the outer layer (crown) generates a deflection effect, while the inner layer's back-to-back inward compression generates a repulsive effect. This mechanism prevents two celestial bodies from colliding when they come close to each other. Additionally, the crown has an additional protective function, as it can deflect incoming asteroids or comets approaching at an angle, effectively redirecting them away from the planet. Concurrently, the crown may capture slow-moving objects and cause them to orbit around the celestial body. This intriguing phenomenon is beautifully illustrated in Saturn's rings, where the ring materials are only within the crown area. This observation suggests that materials placed before the crown would be pushed toward the planet.

Picture this: If Saturn had ten or more rings, it could cover the whole planet and make it look like a much larger planet. Those rings might look like multiple-striped patterns, precisely like the ones we see on Jupiter. Here's why this makes sense. Jupiter's crown area has been collecting materials for billions of years. And its surface is all about orbiting rings, not rotating. The stripe patterns? They're just multiple rings sitting next to each other, each with its orbital speeds. But what about the Great Red Spot? Well, it might be a massive chunk of material hit by an object, crashing into the planet, and getting replaced by the dust to appear in stormy weather. Moreover, the compression of spacetime from above and Undernet the crown creates a dual gravitational system. Making the orbiting crown weightless. This theory offers an alternative understanding of Jupiter's complex, unique features.

When foreign objects, such as meteors, comets, or asteroids, enter the gravitational field, they expand regardless of their atomic compactness due to the increased density of spacetime within this field compared to outer space. This effect becomes more pronounced as the objects approach the Earth's surface, causing further expansion and explosion.

The resistance experienced by these objects while gaining acceleration or entering the gravitational field can be compared to being pushed against an arrow. The increasing density of spacetime as objects approach the surface leads to their explosive demise.

The burning effect observed when dense foreign objects, such as iron-rich asteroids, enter Earth's atmosphere cannot be solely attributed to atmospheric pressure because atmospheric pressure has more of a cooling effect than a burning effect. The Paratheory suggests that the increasing density of spacetime within Earth's gravitational field is crucial in causing these objects to expand and burn.

According to this theory, a spaceship made on Earth may expand when exiting the Earth's gravitational field, potentially compromising structural integrity and amplifying the burning effect upon re-entry.

Elmer Para

Reaching the Size of the Sun: From Localized to Universal Gravity

As the astronaut continues to grow, reaching the size of the sun, they would experience liquefaction from the inside out due to the increasing compression caused by their immense size. However, the compression of spacetime alone is not strong enough to fully liquefy the Sun's matter. There is a specific process at play that contributes to this phenomenon. To better understand this process. Let's revisit the cross-Earth tunnel experiment. Imagine a cross-earth tunnel drawing four bowling balls weighing 10 kg each. According to my theory, these balls would converge at the core with a combined weight of 40 kg as a self-created weight. Now, scale this up to the size of the Sun. Imagine the whole volume of the Sun converging at the parable. This results in simultaneous liquefaction from the inside out.

It's important to note that since spacetime compression passes through all atoms, it destroys atoms from the inside out when spacetime becomes densely concentrated within the parable.

The dynamic interplay between the convergence of matter and parable (double destruction) amplifies the liquefying process and releases pure energy that binds the solar system together.

The equation encapsulates this process:

$$[M + G = E^2]$$

Where:

(M) represents Matter

(G) represents Gravity

(E^2) represents Energy squared

The equation represents the seamless interplay between matter and gravity, producing energy proportional to the square of the combined weight generated by the matter convergence at the parable. This relationship is exemplified by the astronaut's transformation into a shining star.

As Spacetime compression passes through the Sun's core, it weakens the compression, and what remains at the surface is strong enough to contain the Sun's liquid fire in spherical form. (Universal phenomenon.)

Pure Energy

The dynamic interplay between the convergence of matter and the parable of the celestial body generates invisible Pure Energy that binds the solar system together. This pure energy behaves like waves or tentacles on a cosmic scale. The overlapping pure energy-like tentacles interconnecting the solar system as a whole give rise to phenomena such as ocean tides and bulging at the sun's surface.

The active presence of this pure energy can significantly influence the result of the classic double-slit experiment, mainly when conducted with individual subatomic particles. This pure energy causes a light particle to exhibit wave-like behavior, demonstrating that pure energy facilitates directional light particles.

Interconnected Solar System

All planets, including the Sun, experience localized, spherical inward compression of spacetime, known as gravity, which energizes and releases pure energy that binds them together. The Sun, the solar system's centerpiece, functions like a giant octopus with eight tentacles overlapping the energy-like tentacles of the other eight planets, reaching each other, creating a natural bond, and forming an interconnected web solar system.

The motion of planets around the Sun is primarily a result of the momentum generated from the initial alignment of celestial bodies during the solar system's formation. For instance, when all planets are aligned with the Sun, a misaligned planet experiences force from interconnected energy tentacles, pulling it towards the alignment. However, the absence of a braking mechanism leads to overshooting, creating a chain reaction, causing other planets to move similarly, ultimately initiating perpetual momentum and setting the interconnected solar system in cosmic motion.

As Mercury draws closer to the Sun, its gravitational "bubbles" or fields interact, causing Mercury to speed up but remain bound to the Sun due to its binding energy. Extending this interconnected system analogy to all planets in the solar system, any planet deviating from its level will experience restoring force, preserving the overall stability of the celestial system.

The Dynamic Interplay of the Sun and Earth

Focusing on the dynamic connection between Earth and the Sun, they are interconnected through overlapping pure energy tentacles from their respective parables, forming a natural bond reaching each other. These pure energy tentacles extend from Earth and overlap with the Moon, giving rise to high tides on Earth's oceans. Similarly, the Earth's energy tentacles overlap with the Sun's pure energy, causing the bulging of the Sun's surface. As the Earth orbits the Sun, the Earth's parable remains stationary or facing one side to the Sun while the outer layer of the Earth rotates due to the liquified substance around the parable. This phenomenon creates a grinding effect, resulting in the rise of the aurora borealis, and contributes to stability in both rotational and orbital perpetual motion. Moreover, my theory suggests that the lack of rotation in moons and mercury may be attributed to their solidity or lightness. For instance, if the parable of mercury is no longer in the center, gravity cannot reform the planet because of its solidity. In this case, it cannot rotate.

Furthermore, when the parable transforms into a black hole, it acts like a colossal octopus. Its tentacles, extending potent pure energy, bind the whole galaxy together and maintain its flatness and harmony.

Equating black holes as anti-matter

The destructive nature of the celestial body's parable is proportional to its size. Earth's parable destroys atoms from within and releases pure energy. The Sun's parable is even more destructive and generates substantial pure energy that binds the solar system together. While the parable of a star a billion times larger than the Sun begins to solidify and can annihilate atoms, which triggers a supernova explosion, this explosion compresses the parable into the densest form, transforming the parable into a black hole, the pure anti-matter-based on their destructive nature.

A star billion times larger than the Sun: The Explosive Transformation of the Parable into a Black Hole

Our astronaut's journey begins with a transformation of unprecedented proportions. Initially expanding to the size of the Earth and then the Sun, now, at this stage, it has expanded to a billion times larger than the Sun. This immense size induces a colossal spherical inward compression of spacetime, which is gravity itself. Within the center of this massive star lies a solidifying parable, destroying and annihilating atoms, creating a mechanical collision with the inward compression of spacetime or gravity, triggering a supernova event. Imagine, if you can, matter equivalent to a billion times larger than the Sun being compressed by the inward compression of spacetime toward the solidifying parable (becoming an anti-matter core) in spherical form and crushing all matter from the inside out, resulting in both implosion and explosion simultaneously at once. This explosion compresses the parable into its densest form, causing it to metamorphose into a black hole- an entity of pure anti-matter. This awe-inspiring creation can be likened to spacetime in solid state.

The birth of a black hole

The remnants of the supernova's massive explosion gave rise to a "black hole," a pure anti-matter akin to spacetime in a solid form, an entity devoid of atoms, colorless, impenetrable, and indestructible. Its immense density prevents gravity or spacetime inward compression from passing through it, creating an infinite spherical bumping effect. Our astronaut experiences an unfathomable inward compression of spacetime or gravity. They find themselves trapped within their gravitational field where time appears suspended. Any object entering this incredibly dense gravitational field will be eradicated upon contact as if an atom were being pushed against an arrow. Nothing can approach the surface of anti-matter, nor can the contained Anti-matter attack the materials swirling around the crown-like area, creating a delicate balance. This intricate interplay contributes to the beauty and harmony of the cosmos.

The Dance and Sway of Black Holes

If two black holes collide, they merge peacefully, much like liquids, due to their unique properties. However, when one or both black holes are spinning, fueled by their speed, and collide, they can dance, sway, and ultimately escape from the compression of spacetime, resulting in a catastrophic burst that creates a void in space, as we observe today.

As we know it, every galaxy has a massive black hole at its center. Over time, stars within the galaxy will eventually reach the end of their lifespan, leading to supernovas. This supernova may form smaller black holes, which will be drawn and merge with the massive black hole in the center due to their directional attractive force. This process will continue until all the stars die within the galaxy, leaving a supermassive black hole as the sole remnant. And even if all stars vanish within the universe, all these remnants of incredibly supermassive black holes will reign. Their immense mass and directional attractive force will make them appealing to one another, transcending the vast cosmic distances that separate them. This creates a high-speed directional collision that results in catastrophic bursts, marking the end of their existence.

The Final Image

As all atoms vanish, the last two remaining black holes perform an elegant and graceful dance, merging in a mesmerizing cosmic waltz. However, unexpectedly, they break free from the confines of spacetime, causing the cosmic clock to halt its ticking and transform the universe into an infinitely large, formless, dark space frozen in time. The essence of our astronauts is scattered evenly throughout the universe. It may reemerge in different forms for the next cosmic cycle within the infinite self-creating universe as it is conserved.

Section 3

Infinite Self-Creating Universe

Paratheory is rooted in the concept of the infinite self-creating universe, which inherently defies the need to explain a definitive beginning or end. In every proposed finite beginning, a behind always facilitates it, resulting in infinite or endless regress both backward and forward in time.

Space is infinite and cannot be contained by any finite means. Anything that occupies it is governed by the rule of time. Space and time or spacetime form an infinite medium, a flawless and indestructible structure that serves as the creator and container of all things. Finite materials, such as galaxies, stars, planets, and all other observable objects, are merely creations of spacetime. Together, forming the infinite self-creating universe itself.

Illusion of Expansion

Expansion, by definition, does not apply to the infinite universe. However, the endless self-creation of finite materials within this infinite universe creates an illusion of expansion for observers from any location. Galaxies, stars, and planets form and grow gradually and continuously over time rather than all at once, as proposed by the Big Bang theory.

In the Paratheory of gravity theory, Gravity is explained as the simultaneous compression of spacetime in response to the presence of an object. If spacetime were expanding, as the Big Bang theory proposed, we would expect a weakening of spacetime compression or loss of gravity over time since gravity remained constant throughout the universe for billions of years. This is significant evidence against the Big Bang theory, which posits an expanding universe.

Imagine an infinite universe filled with observable objects, including trillions of galaxies, representing only a fraction of the entirety. The Big Bang theory proposes that this vast universe originated from a single point smaller than an atom. This concept is illogical, unscientific, and more akin to magic than reality. In contrast, according to Paratheory, spacetime was infinite even before finite materials evolved. The Big Bang theory can be likened to a small ball held in one's hand. However, an infinite self-creating universe represents the entirety of one's being, encompassing the hand, body, brain, and imagination.

Elmer Para

Genesis of Creation

The universe's creation begins in an infinitely large formless dark space filled with dark matter in perfect balance. In this balance, the dark matter collapses, spacetime comes into play, and creation begins. Spacetime compresses the collapsed dark matter within the almost static environment and becomes a nebula, giving birth to the first generation of colossal stars. As the first generation of stars is born, their energy disturbs the dark matter, transforming it from its static state into a more elusive and dynamic form. This change in dark matter behavior contributes to its widespread presence and ongoing motion throughout the universe. This genesis of creation unfolds throughout the infinite expanse of space. As the first generation of stars reaches the end of their finite lifespans, they give rise to the first generation of primordial black holes. These Black holes amplify creation, acting like cosmic cotton candy machines, attracting dark matter drawing in and coming out as regular matter in the form of dust, a building block of stars. This process leads to the formation and growth of galaxies, which continue to evolve and expand from within, much like a tree branching outward.

Conclusion

As we conclude this book, I want to take a moment to humbly express my gratitude to you, the reader, for embarking on this intellectual exploration. The Paratheory of Gravity is not intended to serve as an absolute answer but rather as an invitation to question, challenge, and expand the boundaries of our understanding.

Science thrives on curiosity, debate, and the willingness to reexamine the foundations upon which its truths rest. I am merely a curious observer of the universe, hoping to contribute a single thread to the vast tapestry of human knowledge. My ideas may not be perfect, and further testing and refinement will be required. However, even the boldest scientific theories begin with a tiny question.

To the aspiring scientists, philosophers, and enthusiasts reading this book, I encourage you to analyze my experiments, scrutinize my claims, and, most importantly, continue asking thought-provoking questions of your own. Progress is a collective effort, and the answers we seek often lie on the other side of rigorous inquiry and open-minded dialogue.

Whether or not the Paratheory of Gravity becomes a stepping-stone for more extraordinary discoveries or sparks fleeting curiosity, I see this as a worthwhile venture. After all, the beauty of science lies in the audacity to imagine a world beyond what we currently know.

Thank you for allowing me to share this perspective with you. It has been a humbling experience to articulate my thoughts and put them forward for consideration. Wherever the pursuit of knowledge leads you next, I hope you carry the same sense of wonder and curiosity that inspired this work.

With deep respect and gratitude, I wish you all the best on your intellectual journey.

Elmer Para

www.ingramcontent.com/pod-product-compliance
Lightning Source LLC
Chambersburg PA
CBHW030101230526
45471CB00003B/1202